AF270700

PHYSICS PROJECT YOUR WAY

Megan Borgert-Spaniol

Super Sandcastle

An Imprint of Abdo Publishing
abdobooks.com

abdobooks.com

Published by Abdo Publishing, a division of ABDO, PO Box 398166, Minneapolis, Minnesota 55439.
Copyright © 2024 by Abdo Consulting Group, Inc. International copyrights reserved in all countries.
No part of this book may be reproduced in any form without written permission from the publisher.
Super SandCastle™ is a trademark and logo of Abdo Publishing.

Printed in the United States of America, North Mankato, Minnesota

102023
012024

Design: Aruna Rangarajan, Mighty Media, Inc.
Production: Mighty Media, Inc.
Editor: Liz Salzmann
Cover Photographs: Adobe Stock; Mighty Media, Inc.
Interior Photographs: Adobe Stock, pp. 1 (scale), 4 (both), 5, 9 (girl), 10, 11 (car, friction graphic), 15, 24, 28 (car, friction graphic), 31; iStockphoto, pp. 6, 14 (stool, bubble wrap), 21 (bubble wrap), 26, 27, 29 (bubble wrap); Mighty Media, Inc., pp. 14 (board, ruler, car, tape, felt), 16 (experiment), 17 (experiment), 18 (experiment), 21 (board, felt), 28 (experiment), 29 (board, felt); Shutterstock, pp. 7, 8, 9 (boy), 11 (boy), 13, 14 (stopwatch), 19, 22, 25, 29 (boy), 30
Design Elements: Shutterstock

Library of Congress Control Number: 2023939347

Publisher's Cataloging-in-Publication Data
Names: Borgert-Spaniol, Megan, author.
Title: Physics project your way / by Megan Borgert-Spaniol
Description: Minneapolis, Minnesota : Abdo Publishing, 2024 | Series: DIY science fair fun! | Includes online resources and index.
Identifiers: ISBN 9781098292072 (lib. bdg.) | ISBN 9781098278977 (ebook)
Subjects: LCSH: Do-it-yourself work--Juvenile literature. | Physics--Juvenile literature. | Science projects--Juvenile literature. | Science fair projects--Juvenile literature.
Classification: DDC 507.8--dc23

Super SandCastle™ books are created by a team of professional educators, reading specialists, and content developers around five essential components—phonemic awareness, phonics, vocabulary, text comprehension, and fluency—to assist young readers as they develop reading skills and strategies and increase their general knowledge. All books are written, reviewed, and leveled for guided reading, early reading intervention, and Accelerated Reader™ programs for use in shared, guided, and independent reading and writing activities to support a balanced approach to literacy instruction.

CONTENTS

EXPLORE PHYSICS

Do you love to ride roller coasters? Do you wonder how planes fly? You might enjoy physics! Physics is the study of **matter** and its motion. Scientists who study physics are called physicists.

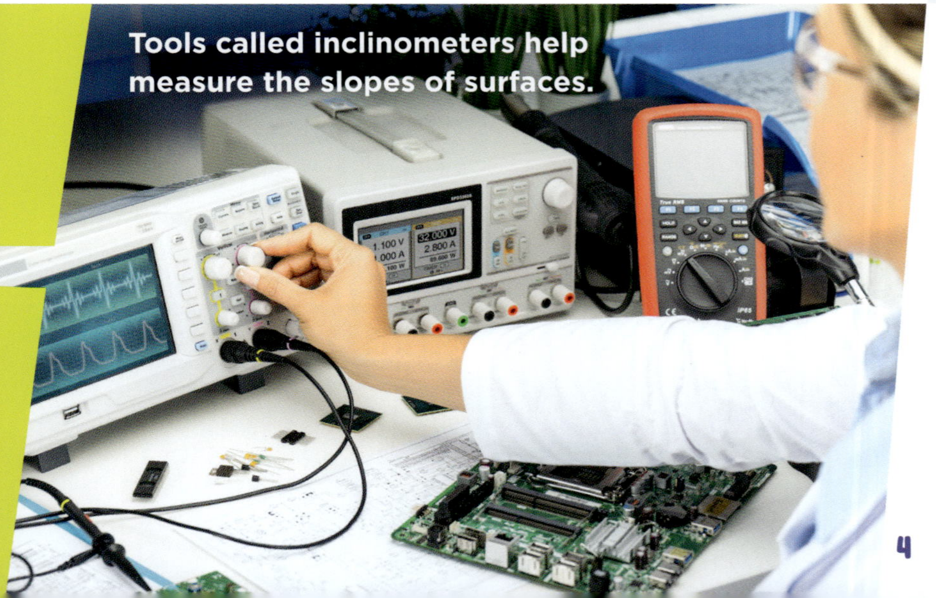

Physicists study how energy and forces affect matter. Types of energy include heat and sound. Examples of forces include **friction** and gravity.

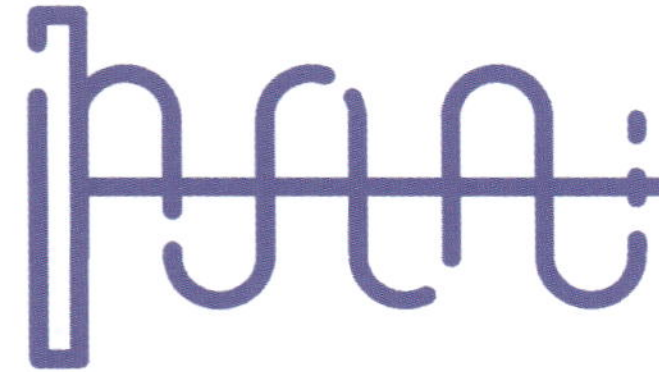

Physicists use scales and balances to measure the weight and mass of objects.

BECOME A SCIENTIST!

Scientists use a process called the scientific method. Check out the steps on the next page. You will use this method to **design** your own physics project!

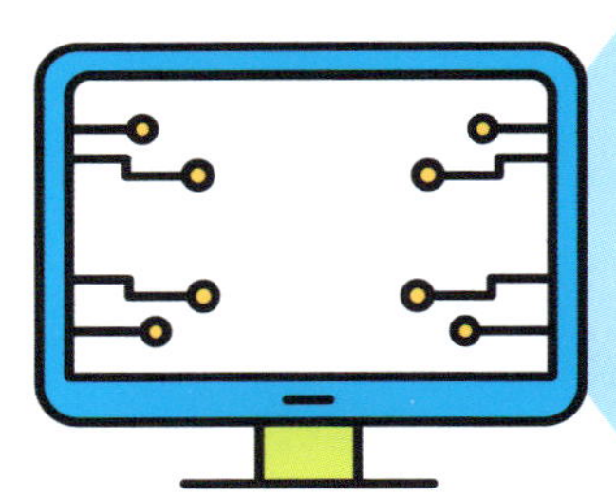

THE SCIENTIFIC METHOD

1 ASK A QUESTION
What would you like to find out?

2 GATHER INFORMATION
What information do you need to understand your topic?

3 FORM A HYPOTHESIS
What do you think is the answer to your question?

4 EXPERIMENT
How can you test your hypothesis to find out if it is correct?

5 RECORD THE RESULTS
What did you observe in your experiment?

6 WRITE A CONCLUSION
Did your results support your hypothesis?

ASK A QUESTION

What topic do you want to learn about?

Maybe you are interested in sound energy. Or maybe you want to know more about **friction**. Start asking questions! Write your questions in a notebook so you don't forget them.

8

What is friction?
How does friction affect motion?

GATHER INFORMATION

Scientists often have many questions they'd like to answer. But for now, choose one to **focus** on. Save the others for **future** projects.

It's time to **research** your **topic**. You can gather **information** from many different sources.

Read online articles about the topic.

Read books about the topic.

Talk to scientists or other experts.

What did you learn about your **topic**? Write it down in your notebook. Then you'll have all the **information** you need in one place.

FORM A HYPOTHESIS

After you research your topic, it's time to form a hypothesis.

Your hypothesis is what you believe is the answer to your question. First, revisit your question. Do you want to change it based on what you learned? Then think of a few different hypotheses. Record them all in your notebook.

QUESTION: How does **friction** affect the speed of a moving object?

HYPOTHESIS 1

An increase in friction results in an increase in speed.

HYPOTHESIS 2

An increase in friction results in a decrease in speed.

HYPOTHESIS 3

Friction does not affect the speed of a moving object.

I learned that **friction** works against motion.

Now, choose which hypothesis makes the most sense based on your **research**.

So, I think an increase in friction results in a decrease in speed.

13

PREPARE YOUR LAB

Find an area with a sturdy table or counter to work on. Then gather the supplies you'll need for your science experiment.

SUPPLIES

board about 28 by 10 inches (71 by 25 cm)

low stool or bench

ruler

toy car

stopwatch

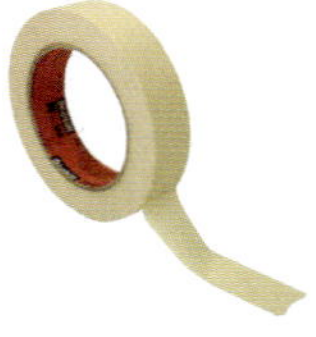

tape

felt

bubble wrap

All labs have rules that scientists have to follow. Here are some rules for your lab. They will help you stay safe and have fun while doing your experiment!

➡ **Ask an adult** for permission to use the materials and do the experiment.

➡ **Ask for help** with sharp or hot tools.

➡ **Wear goggles** and gloves to protect your eyes and hands.

➡ **Clean up** when you are done and put everything away.

EXPERIMENT!

1 Set one end of the board on the ground. Lean the other end against the stool. Have a helper hold the toy car so its back wheels are lined up with the top end of the plank.

2 Tell your helper to let go of the car. Start the stopwatch at the same time. Stop the stopwatch the moment the car hits the ground. Record the time in your notebook.

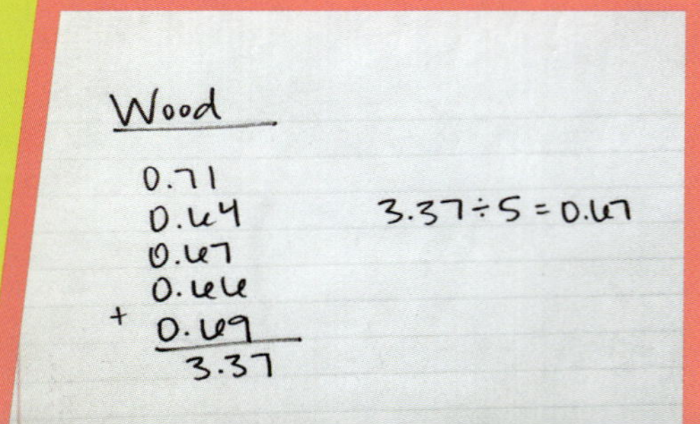

3 Roll the car four more times. Add up the five recorded times and divide that number by five. This gives you the average time it took the car to roll down the track.

4 Tape felt to the top surface of the board. Let the car roll down it five times. Record how long it took each time. Calculate the average.

5 Replace the felt with bubble wrap. Let the car roll down the board five times. Record how long it took each time. Calculate the average.

LAB TIP

Make sure the slope of the board stays the same throughout the experiment.

Look at the results of your experiment so far. You might be ready to draw a conclusion. But first, consider any other **variables** that might affect your results.

The slope of the wood plank affects how fast the car moves. You kept the slope the same for all three tests.

Keeping the slope the same means slope was not a variable in this experiment.

The only **variable** in the experiment was the roughness of the track. The rougher the track, the more **friction** was created.

RECORD THE RESULTS

During experiments, scientists record data and other observations. You wrote down how long it took the toy car to travel down the track with each surface. Now, it's time to record your data to share with others.

Scientists often use tables and graphs. This helps make the results easy for others to understand.

A table organizes **information** in rows and columns.

TRACK MATERIAL	FRICTION RANKING (number increases with roughness)	AVERAGE TIME TO BOTTOM (seconds)
Wood	1	0.67
Felt	2	0.77
Bubble Wrap	3	0.79

A line graph shows the relationship between two **variables**.

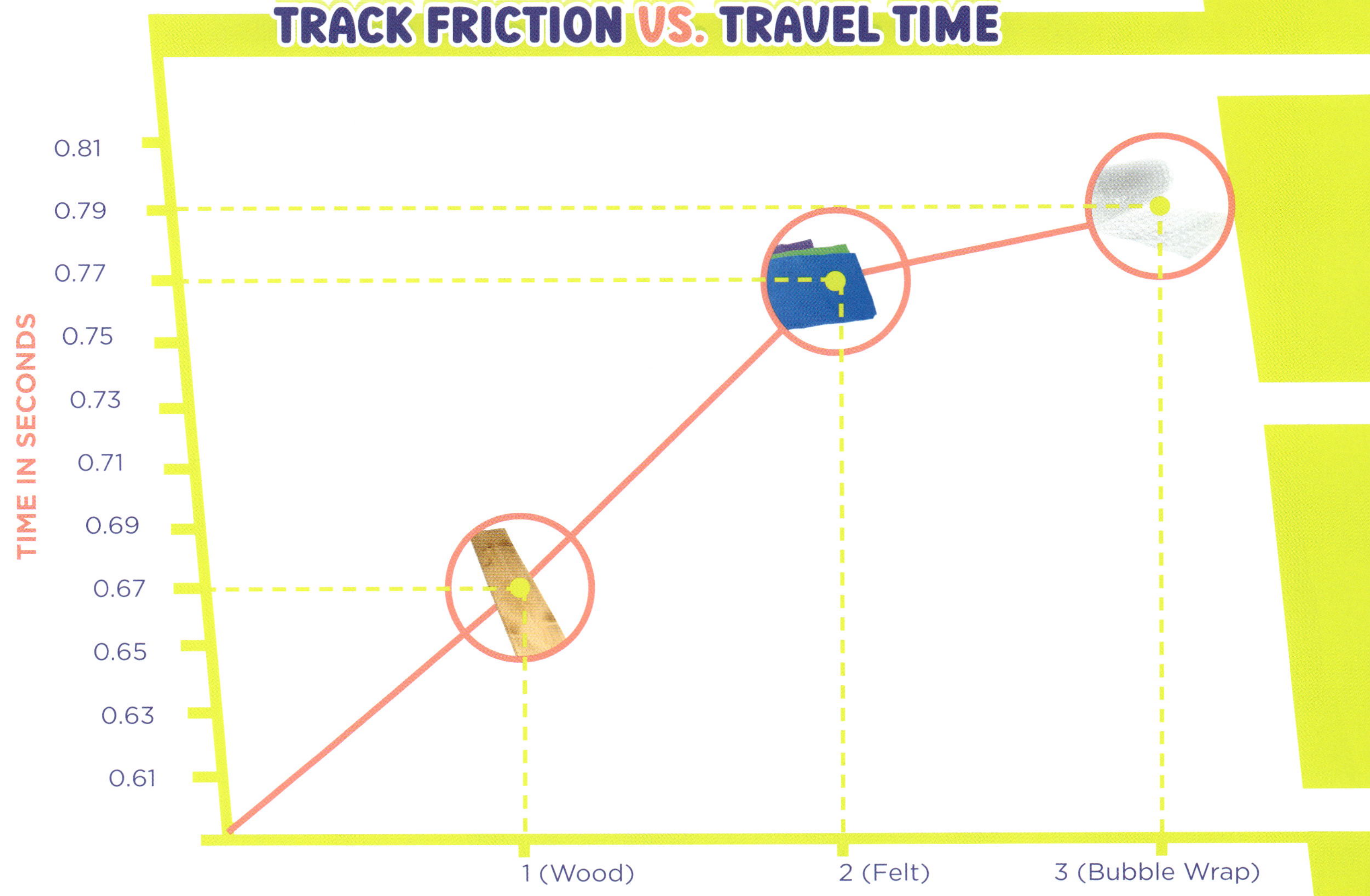

WRITE A CONCLUSION

You recorded your results. Now it's time to write your conclusion. This is a **summary** of your experiment. Your conclusion provides the answer to your original question. It also states whether your results support your hypothesis.

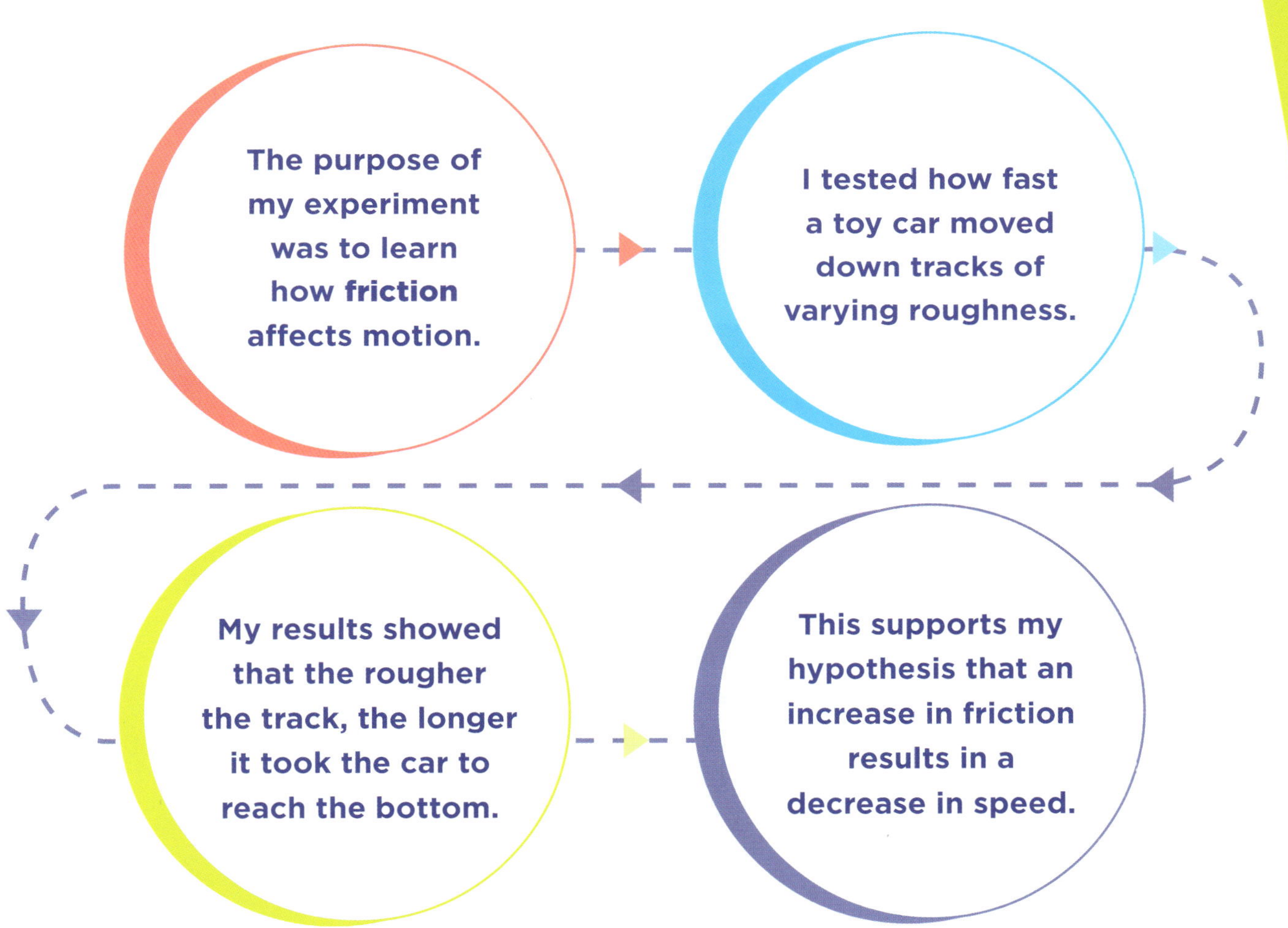

Many scientists find that their hypotheses were wrong. There is nothing wrong with being wrong! It means you did your experiment without **bias** and were surprised by the results. That is another mark of a great scientist!

FURTHER RESEARCH

A conclusion offers an answer to your original question. But it can bring up new questions too! For scientists, the end of one experiment often leads to more **research** and new hypotheses to be tested.

What new questions do you have? What could you research further? Do you have a new hypothesis to test?

Does water affect the friction between two objects?
How does friction affect wind speed?
Does the mass of an object affect its speed?
25

PRESENT YOUR PROJECT

Young scientists share their **research** at science fairs. Students share what they learned with classmates, teachers, parents, and sometimes judges. It is a chance to show all the work they put into their experiments.